MATHEMATICS BEHIND THE SCENES

CONTENTS

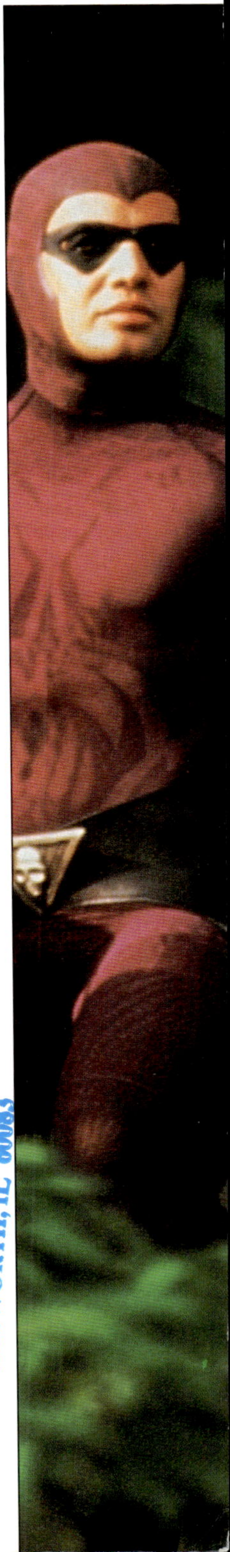

JAMES BURNETT
CALVIN IRONS

Movie Magic

For more than a century, the movies have captivated audiences with action, drama, suspense, and intrigue. However, a true appreciation of cinema begins with a glimpse behind the scenes – to see how the huge movie-making business really works.

Movie Sets

The movie *The Fall of the Roman Empire* had the biggest movie set ever built. It covered an area measuring 1,312 feet by 754 feet, and took 1,100 people six months to complete. Before the set was constructed, the designers made detailed plans, drawings, and models.

Did you know?
More than one-and-a-half miles of film are used for a full-length movie – that's 90 feet of film for each minute of viewing.

Many countries have awards and festivals to celebrate each year's greatest movie achievements. The most famous prizes are the Academy Awards, presented each year in Hollywood. The first Academy Awards ceremony took place in 1929.

Some movies make huge profits. *Jurassic Park,* which cost $60 million to produce, earned more than $900 million in its first two years on the screen.

Fabulous Flicks

When we see movement on the big screen, we are actually seeing thousands of still pictures. Each picture or *frame* shown is only slightly different from the previous one. When the frames flick very rapidly from one to the next, we see them as one continuous moving image.

Moving pictures were used long before photography was invented. When this old "picture wheel" is spun around, it creates an illusion of people moving.

Eadweard Muybridge was the first person to try capturing movement in a series of photographs. In 1878, he took these pictures of a woman jumping.

The first *movies* were shown on *Kinetoscope* machines like the ones shown in the picture. Each movie took about 15 seconds and required 40 feet of film. A ticket cost five cents.

Quick Flicks

This camera-projector was invented in France in 1895 by the Lumière brothers. To project a film, the operator had to turn the handle twice every second. Fifteen frames were shown per second.

The Rough and the Smooth

Most modern movie cameras capture 24 frames per second – enough to create the illusion of a "smooth" moving image. Early movie cameras captured fewer than 20 frames each second, which created a slightly jerky or flickering effect. This is why early moving pictures were nicknamed *flickers* or *flicks*.

1870 1880 1890 1900 1910 1920 1930

Fascinating Facts

Movie History

1906: The world's first full-length movie, *The Story of the Kelly Gang,* was produced in Melbourne, Australia. It was approximately 60 minutes long.

1927: The first full-length movie made with sound, *The Jazz Singer,* was released. Sound movies were later nicknamed *talkies*.

1903: The first movie to tell a story was *The Great Train Robbery*. It was only 11 minutes long.

1929: Silent movies quickly disappeared. *The Poor Millionaire* was the last to be released.

Technology and Innovation

1889: George Eastman invented flexible film for Kodak cameras. Later, Edison used this type of film.

1932: The Technicolor Company produced a camera designed to film in color.

1878: Eadweard Muybridge was the first person to photograph movement.

1891: Thomas Edison invented the *Kinetoscope,* a machine used to view movies.

1911: G.A. Smith invented the first effective color photography process known as *Kinemacolor.*

1895: The Lumière brothers invented a machine that could be used both to make and show movies.

1870 1880 1890 1900 1910 1920 1930

1940 1950 1960 1970 1980 1990 2000

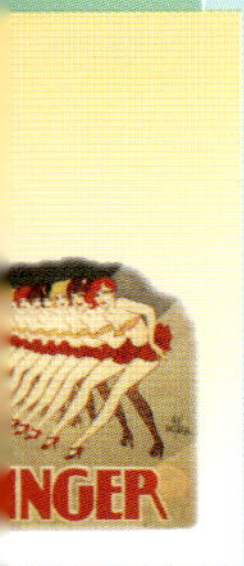

Between 1946 and 1956, as television became popular, attendance at theaters in the United States was almost cut in half.

1937: Disney's *Snow White and the Seven Dwarfs* was the first full-length animated movie to be released.

1952: *3-D movies* were screened for the first time. The audience wore special glasses to see the 3-dimensional effect.

1964: *The Fall of the Roman Empire* was filmed on the largest set ever built.

1982: *Gandhi,* the movie with the largest "cast" ever, was filmed. It is believed that there were more than 300,000 *extras* in one scene.

1994: *True Lies,* with Arnold Schwarzenegger, was the first movie to have a budget greater than $100 million.

1. How many years ago was:
   a. the first full-length movie released?
   b. the first full-length animated movie released?

2. In which decade were *3-D movies* introduced to audiences?

3. How many years after the first full-length *talkie* was shown was the last silent movie released?

4. Name three technical innovations from the 19th century.

Look at the information on pages 4–5.

5. How many frames would there be for one minute of film using:
   a. the Lumière brothers' machine?
   b. a modern movie camera?

1940 1950 1960 1970 1980 1990 2000

Making Movies

Filming the action of a movie is only one stage of the movie-making process. Planning and organization begin well before the start of filming, while other important tasks are carried out after most filming is complete.

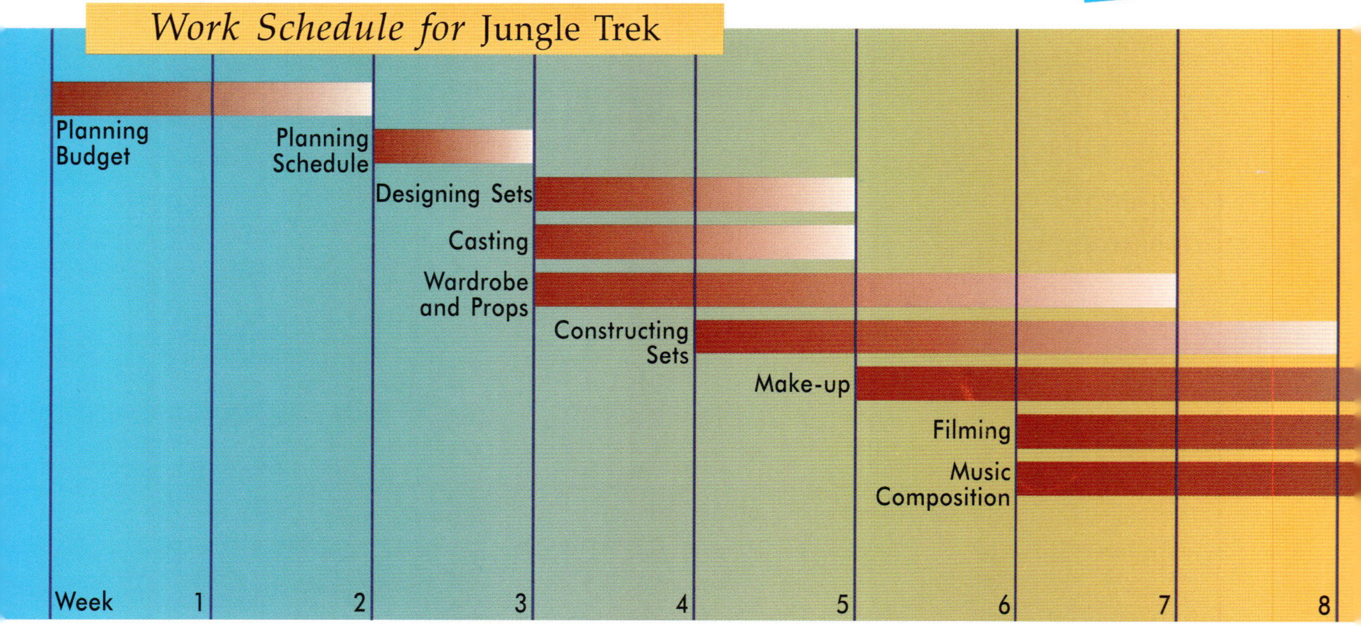

A work schedule is an important planning tool. It shows when different tasks should take place.

Sometimes a movie-maker is both the producer *and* director, managing the business side and the creative side of a movie. Steven Spielberg has produced and directed six of the top 20 box-office successes of all time.

The work schedule below is a *guide* for how much time might be spent on various jobs, but each task shown on the schedule also requires more detailed planning. This filming schedule for the movie *E.T.* shows the time allocated for filming at different locations.

Many tasks overlap during movie production. For example, while some scenes are being filmed, the sets for others might still be under construction.

E.T. FILMING SCHEDULE
TOTAL DAYS FILMING: 61

2 days	on location	school scenes
11 days	on location	exterior scenes Northridge, Tujunga
42 days	studio	interior house scenes backyard and hillside Redwood clearing and landing site
6 days	on location	exterior scenes small town

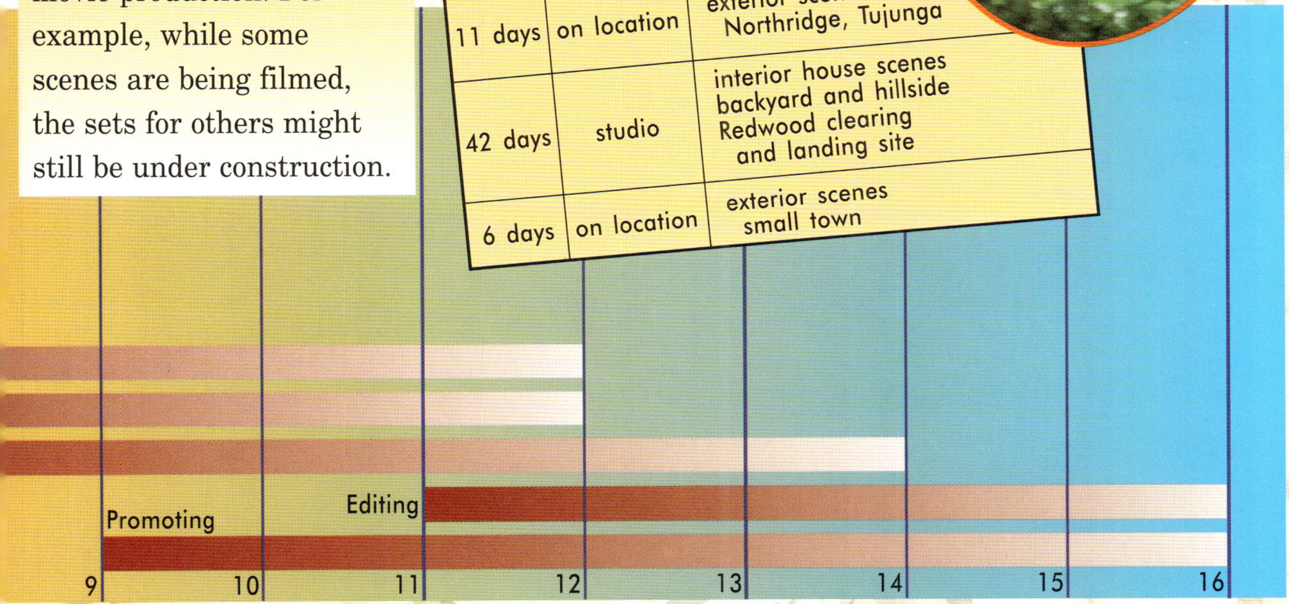

Around the Clock

Most of the day-to-day organization and decision making is handled by the production manager. The 24-hour time line opposite shows what a production manager might do during a busy day.

1. How many hours does the production manager spend at work:

a. during this typical day?

b. during a typical 5-day week?

2. What fraction of a day does the production manager spend:

a. sleeping? **b.** on set? **c.** in meetings?

Look at the *Jungle Trek* work schedule on pages 8–9.

3. a. Which task took the longest time? . . . the shortest time?

b. The carpenters constructing sets worked a 6-day week. How many days did they take to complete their task?

c. The camera operators worked a 5-day week. Each day they worked 10 hours. How many hours did one operator work during the filming of *Jungle Trek*?

4. Look at the *E.T.* filming schedule on page 9. If the movie crew worked a 5-day week, how many weeks did filming take?

The production manager has a 5 o'clock meeting with the set designers and art department.

Big Budgets

For a "typical" Hollywood movie, a producer needs to raise about 20 million dollars. For a big-budget production, the cost of making a movie can be more than 100 million dollars! The pie graph gives you a picture of the costs for a big-budget movie.

Where does it all go?
Money is required for every aspect of making a movie, from paying the actors to providing meals for the cast and crew when filming on location.

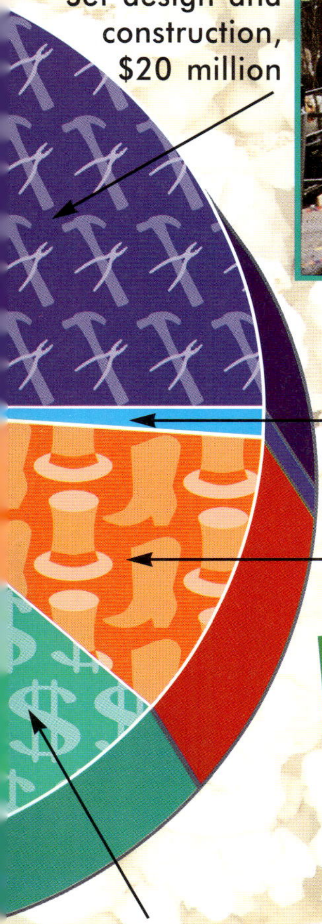

Did you know?
The first actor to earn more than one million dollars for a single movie was Elizabeth Taylor, for her starring role in the 1963 epic *Cleopatra*. Today's highest-paid Hollywood stars earn as much as $20 million per movie.

MONEY, MONEY, MONEY

The producer of a movie is responsible for creating the budget and making sure it is followed. Once the budget has been created, the producer compares the actual cost for each item with what was listed in the budget.

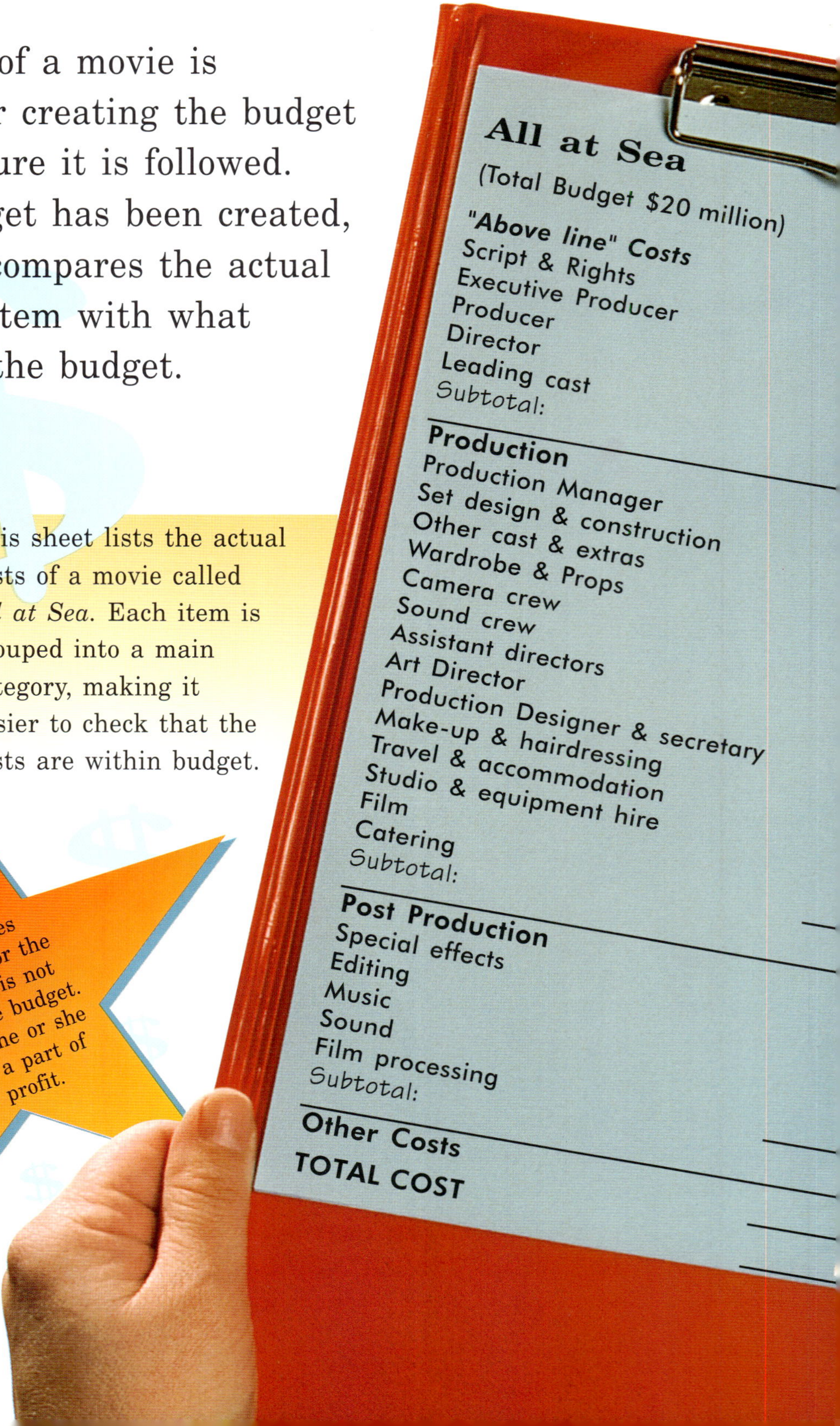

This sheet lists the actual costs of a movie called *All at Sea*. Each item is grouped into a main category, making it easier to check that the costs are within budget.

Sometimes payment for the producer is not part of the budget. Instead, he or she is paid a part of the profit.

A big budget doesn't guarantee a movie's success. *Last Action Hero* cost more than $120 million to make, but its box-office earnings were only $44 million.

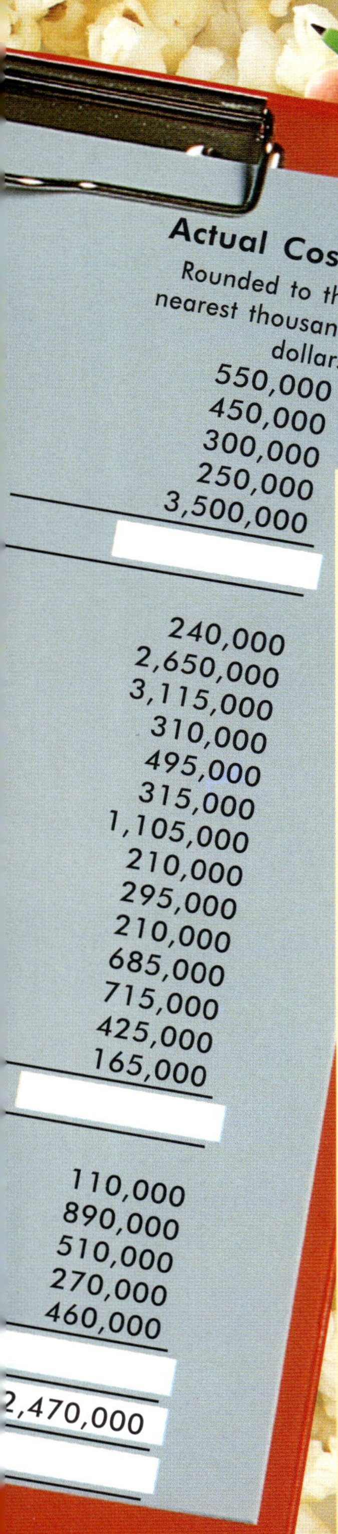

1. Estimate the total cost of *All at Sea*. Was it within the $20-million budget?

2. List in order the 10 most expensive items.

3. Which item was:

a. closest to half a million dollars?

b. closest to one million dollars?

4. Calculate each subtotal and the total. About what fraction of the total cost was:

a. the production category?

b. the post-production category?

Look at the pie graph on pages 12–13.

5. What was the total budget?

6. What fraction of the total was each budget category?

7. Find two categories which made up about:

a. one-half of the budget

b. one-fourth of the budget

c. one-tenth of the budget.

A Day's Work

A day at work during filming might last from early in the morning until late at night. Not all members of the cast and crew work nonstop during this time, but people from every department are likely to be on the set for very long hours.

Helping actors to "become" their characters is the main task of the make-up department. This can take many hours each day.

Actual filming might start at 8 A.M., but some people start work much earlier and finish long after filming is over. The clocks on this page show some possible starting and finishing times.

The unit department crew is responsible for setting up and tearing down everything needed for a day's shoot. Because they begin early, they take a 4-hour break in the middle of the day.

Production Manager

Make-up Department

The team building the sets works long hours to make sure that everything will be ready for filming in a few days' time.

Set Construction Crew

ON THE SCENE

The day-to-day work of actors can vary greatly. Sometimes they need to appear in many scenes but on other days they may not be filmed at all!

Whoopi Goldberg enjoys a break between acting in scenes from *Eddie*.

Look at the 3-day shooting schedule on page 19.

1. Which day:

a. has the most scenes?

b. covers the most pages of the script?

2. How many scenes does Character 1 appear in? How many pages of the script are in those scenes?

3. What was the total number of pages of the script filmed at night?

Look at pages 16–17.

4. Which crew members work:

a. more than 12 hours a day?

b. less than 12 hours a day?

5. Calculate the total number of hours that a member in each of these departments might work in a 5-day week:

a. Make-up department

b. Camera crew

c. Unit department.

Scene Number	Interior/ Exterior	Location	Summary of scene	Day/ Night	Page Count	Character Numbers
Day 1						
19	Exterior	At Sea	On the deck	Day	$\frac{1}{8}$	1, 2, 3, 4, 6, 7
21	Exterior	At Sea	Laughing over a joke	Day	$1\frac{6}{8}$	1, 2, 3, 4, 6, 7
20	Exterior	At Sea	Studying the maps	Day	$\frac{7}{8}$	14, 15
24	Exterior	At Sea	Patrol boat along side	Day	$\frac{7}{8}$	1, 3, 4, 8, 9, 11
30	Exterior	At Sea	Officer questions Lisa	Day	$1\frac{3}{8}$	14, 17
2	Exterior	At Sea	Just before sunrise	Night	$\frac{1}{8}$	
6	Exterior	At Sea	Coming quietly ashore	Night	$1\frac{2}{8}$	14, 28
Day 2						
72	Interior	Studio	Lucas surprises Jake	Day	$\frac{6}{8}$	1, 13
74	Interior	Studio	There's a scuffle	Day	$\frac{2}{8}$	1, 9, 13, 15
10	Interior	Studio	Jeremy starts the boat	Day	$\frac{3}{8}$	3
12	Interior	Studio	Calling for direction	Day	$1\frac{7}{8}$	2, 10
18	Interior	Studio	Up the stairs	Night	$\frac{1}{8}$	9, 11, 16
53	Interior	Studio	A frantic phone call	Night	$1\frac{2}{8}$	4, 8, 11
60	Interior	Studio	Smashing down the door	Night	$\frac{4}{8}$	7, 14, 15
Day 3						
9	Exterior	House	Leaving the villa	Day	$\frac{2}{8}$	1, 2, 3, 4, 6, 7
5	Exterior	House	Breakfast on the balcony	Day	$\frac{4}{8}$	1, 2, 3, 4, 6
7	Exterior	House	A heated discussion	Day	$2\frac{3}{8}$	1, 2, 4, 8, 9
28	Interior	Studio	Going below	Day	$\frac{1}{8}$	1, 14, 15
27	Interior	Studio	Counting the cash	Day	$\frac{3}{8}$	1
81	Interior	Studio	Checking the news	Day	$\frac{6}{8}$	4, 11
67	Interior	Studio	Pauline looks on	Night	$\frac{1}{8}$	9
68	Interior	Studio	Pauline helps out	Night	$\frac{2}{8}$	1, 13
76	Interior	Studio	Police at the door	Night	$\frac{4}{8}$	20, 25
77	Interior	Studio	Lucas and Jake alarmed	Night	$\frac{4}{8}$	1,13

Each day's work during filming is built around a *shooting schedule.* The characters are listed on the schedule by numbers rather than by names.

Each scene covers one section of the script – sometimes only a few lines. Scenes are "measured" in eighths of a page, and this information is carefully recorded on the shooting schedule.

Cast of Thousands

Matching the right actor to a part is vital for the success of a movie. Sometimes a famous *star* is invited to play a lead role, but often hundreds of actors are considered before a choice is made.

Sometimes a part calls for an actor with a "special look." The movie *Funny Face* needed an actor who looked like the star of the movie, Audrey Hepburn.

Sometimes the main character is an animal. In the movie *Babe,* 48 pigs were used for the lead role. If the makers had used only one, it would have grown far too big for the part before the end of filming. Many other animals were used in the movie. Altogether there were more than 500 animals, and 56 animal handlers.

Most actors have to audition for a role. They read from a script or act in a short scene. Only a few make it to the last stage when the final cast is selected.

Extra! Extra!

MOVIE	MAIN ACTORS	EXTRAS
The Last Emperor	60	19,000
Gallipoli	43	4,000
Gandhi	145	300,000

Many of the people who appear in movies have no lines to say, and are seen only for a few seconds or as part of a crowd. These people are called *extras*. As the chart and the photo from the movie *Gandhi* show, some movies need extras by the thousands!

People Puzzles

A movie often brings together a wide range of actors – young, old, male, female, stars, and extras. The diagram below shows that an actor can belong to more than one category.

Actors in Family Holiday

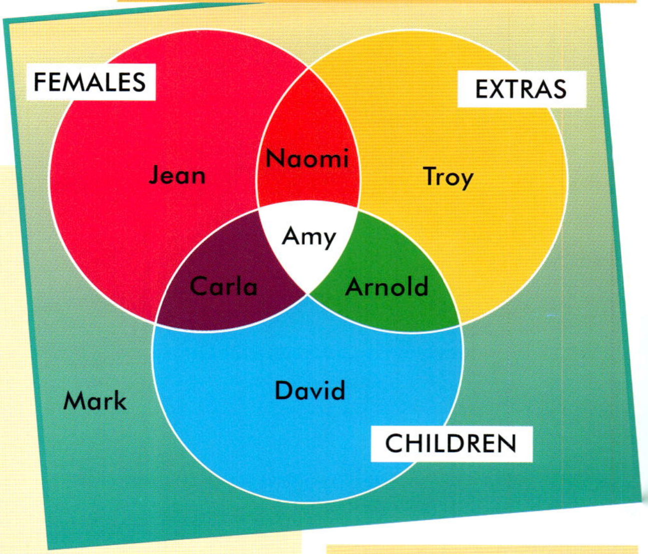

Look at the diagram.

1. Who do you think is:

a. a male child actor who is an extra?

b. an adult female who has a leading role?

2. How many female extras are in the cast of *Family Holiday*?

3. Write all you know about these actors.

a. Carla **b.** Amy **c.** Troy

4. How would you describe Mark?

5. Copy the chart on page 23 and mark the correct box for each character.

6. Make up your own "who's who" clues about an imaginary movie.

Look at the Cast of Thousands chart on page 21.

7. Suppose the *extras* receive $50 per day. How much would the producer pay for the extras in each of the three movies:

a. for one day? **b.** for 5 days?

Venn diagrams use overlapping circles to help sort and display information. Each circle represents a different category. If an item is shown in three different circles, it belongs to three categories.

Who's Who in All at Sea?

Six actors each have a part in the movie *All at Sea*, which is set on a cruise ship. Use the clues and the chart below, to help you figure out which part each actor plays.

A. Chris plays the character who is in charge of the voyage.

B. Shane plays an assistant.

C. Sam plays a character who helps to keep people healthy.

D. Terry's character does not work directly with any of the other characters.

E. Jean will need to be a good swimmer.

F. Manuel works closely with Chris.

Roles

Actors	Life Guard	Doctor	Medical Secretary	Cook	First Mate	Captain
Chris						
Shane						
Sam						
Terry						
Jean						
Manuel						

Logic puzzles were made popular by an English mathematician called Charles Dodgson, who was also known as Lewis Carroll, the author of *Alice in Wonderland*.

Scale Models

Sometimes it is not practical to film things that are "life-size," or as big as they will appear on screen. Instead, small-scale replicas are used. They are filmed in a way that creates an *illusion* of greater size.

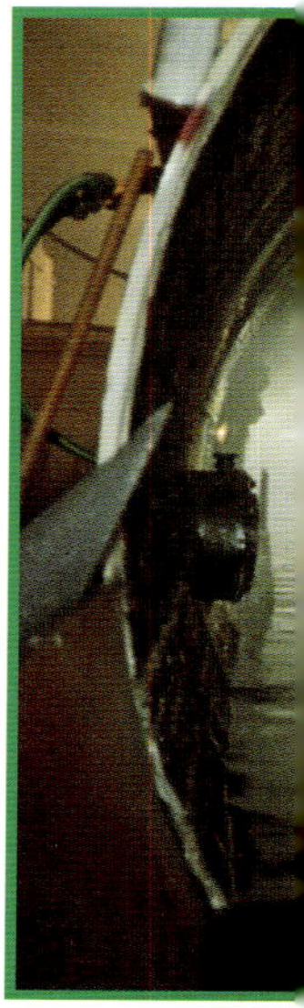

Models of various sizes were used in *The Empire Strikes Back*. The *desert walker* that is closest to us in the scene above looks about 10 stories high. But actually it is only as tall as a 10-year-old child. The other models are even smaller!

The producers of *Batman Returns* saved time and money by using this model vehicle. In the movie it looks large enough to carry Batman through the water tunnel.

Larger than Life

Sometimes scale models need to be *larger* than life-size. Props such as a pair of enormous glasses constructed for the movie *Hook* made Tinkerbell appear to be only seven inches tall.

Sometimes model-makers construct whole towns, carefully designed to appear real when filmed. This technique is especially useful in disaster movies, which often require buildings to be destroyed.

Precise Plans

Many things need to be built for movies, from scale models to full-size buildings. Before a building is constructed, the designer prepares detailed scale drawings and plans to show its exact dimensions.

Scale drawings were important in the design and construction of this model for *Star Trek*.

Look at the building plan.

1. a. Calculate the length and width of the building:
- including the porch
- not including the porch.

b. Write the length and width of each of the three rooms inside the building.

c. Find the floor area of the building, not including the porch. How does this compare to the area of your classroom?

2. Look at the side view. Estimate the height of the front door and of the side window. Describe how you made your estimates.

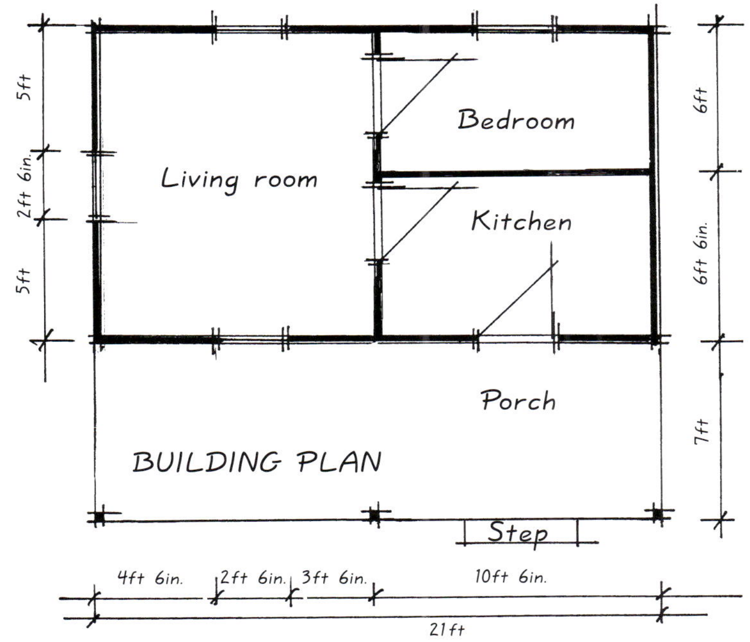

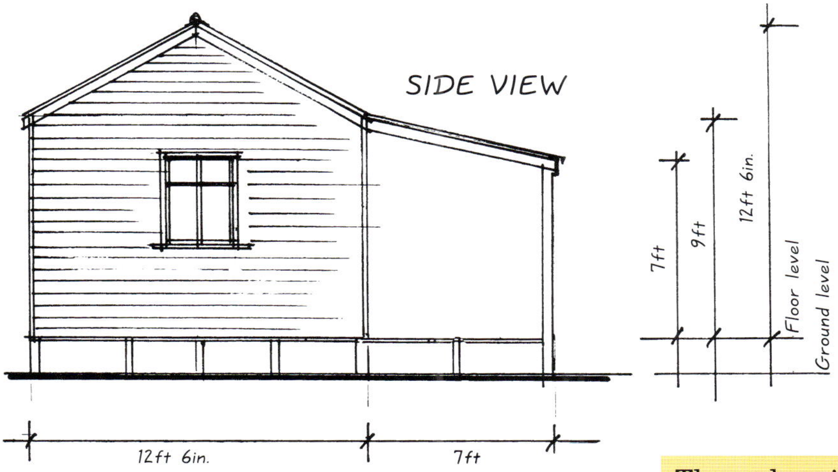

These drawings show a building that was constructed for a movie about an early gold-mining town.

Movie Cameras

All movie cameras record images on moving film, but the cameras themselves vary greatly. Throughout movie history, cameras have been designed with many different features, depending on the technology of the time, and also to suit particular purposes.

Bell and Howell

Some early movie cameras were designed to hold an extra film spool, giving the camera "Mickey Mouse ears." With the extra spool attached, this camera, made in the 1930s, weighs 13 pounds.

Debrie Le Parvo

Until the 1930s, most cameras were hand-cranked, which meant that the operator had to turn a handle to keep the film in the camera moving. This hand-cranked camera, made in France in 1922, weighs 24 pounds.

Classic Mitchell

Weighing 132 pounds, this is one of the largest and heaviest cameras ever used. The first Mitchell camera was made in the 1920s, and these cameras were in common use until the 1960s.

Many modern cameras are lightweight, but this underwater camera weighs a heavy 81 pounds. It is designed so that it does not sink or rise when it is under water. This helps the operator to hold it steady.

ON A ROLL

Two different measurement systems are used to measure movie film. The length of film is measured in *feet*, while film width is measured in *millimeters*.

Several different film widths are used in movie-making. These are shown at actual size. The most common is 35-millimeter film which is used in making most modern feature movies.

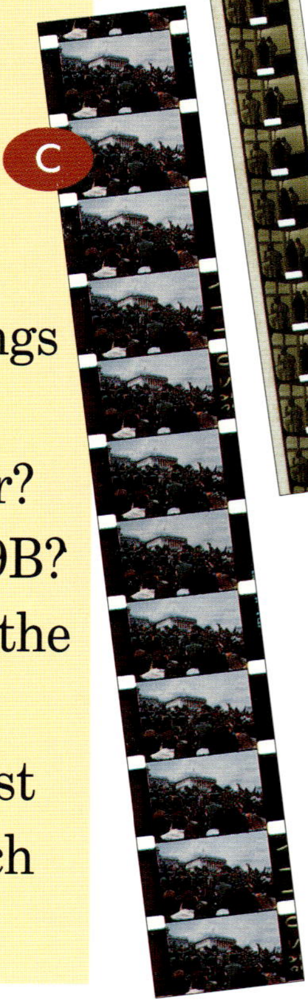

1. Measure the width of each film strip. Make a chart to show how much wider strip A is than the other four strips.

2. Look at the camera sheet on page 31.

a. How would you use the counter readings to calculate the length of each take?

b. How much film was printed altogether?

c. How much film was taken for scene 19B?

d. What fraction of the 400-foot roll was the film taken for 19B?

3. Look at pages 28–29 and find the heaviest camera. How much heavier is it than each of the other cameras?

Camera Sheet

TITLE:	Orange Moon
DATE:	July 31 (Camera A)
ROLL NUMBER:	1
LENGTH OF ROLL:	400 feet

SCENE NUMBER	TAKE NUMBER	COUNTER READING (at end of take)	TAKE LENGTH (feet)	PRINT
15A	1	35	35	
	2	95	60	P
19A	1	115	20	
	2	140	25	
	3	150	10	
	4	170	20	
	5	190	20	
	6	200	10	P
	7	230	30	P
19B	1	275	45	
	2	290	15	
	3	310	20	P
19C	1	320	10	
	2	335	15	
	3	345	10	P
19D	1	385	40	P

TOTAL PRINTED

UNUSED AT END

TOTAL WASTE

"Take 1..."

Usually a scene has to be filmed more than once before the director is satisfied. Each filming of a scene is known as a *take*. A camera sheet listing all the takes is filled out as filming proceeds. This provides a record of every foot of film.

P for Print

The letter "P" is written in the "Print" column when the director decides that a take is good enough to be printed.

Counter Reading

A counter on the camera shows how much film has been used. This is recorded in the "Counter Reading" column.

Amazing Animation

In most movies we see actors on the screen, but some movies feature cartoon characters or figures made from clay. These characters are brought to life by a technique known as *animation*.

Animation artists break every action (such as a step or jump) into several stages. These separate pictures blend together in the final movie, creating an illusion of movement.

Cel Animation

Cel animation has been used for many years. It involves drawing thousands of pictures, called *cels.* Each cel is photographed individually by a rostrum camera like the one shown in the picture. Each cel becomes a frame on the film.

Clay Animation

Animated movies can be made using clay models. Different frames show the models in slightly different positions. When the movie is shown, the figures appear to move by themselves.

Computer Animation

Computers can be used to create very complex animation effects. For example, they can combine cartoon figures with real-life characters.

TIME (SECONDS)	NUMBER OF FRAMES	NUMBER OF FEET
1	24	1½
2	48	3
3	72	4½
4	96	6
5	120	7½
6	144	9
7	168	10½
8	192	12
9	216	13½
10	240	15

Just one second of animated film requires 24 frames. This means that a feature-length cartoon could be made up of as many as 120,000 frames. You can use the chart above to calculate the number of frames and amount of film strip needed for a feature-length cartoon.

Step by Step

Each second of a cartoon contains 24 frames. If the animator wants to show a character running fast, each step will be completed in just a few frames. But to make a character walk slowly, many frames are needed to show a single step.

Each set of frames below shows a character taking one "step." In set A, the character completes one step in four frames. So in one second of film (24 frames) the character in set A will take six steps.

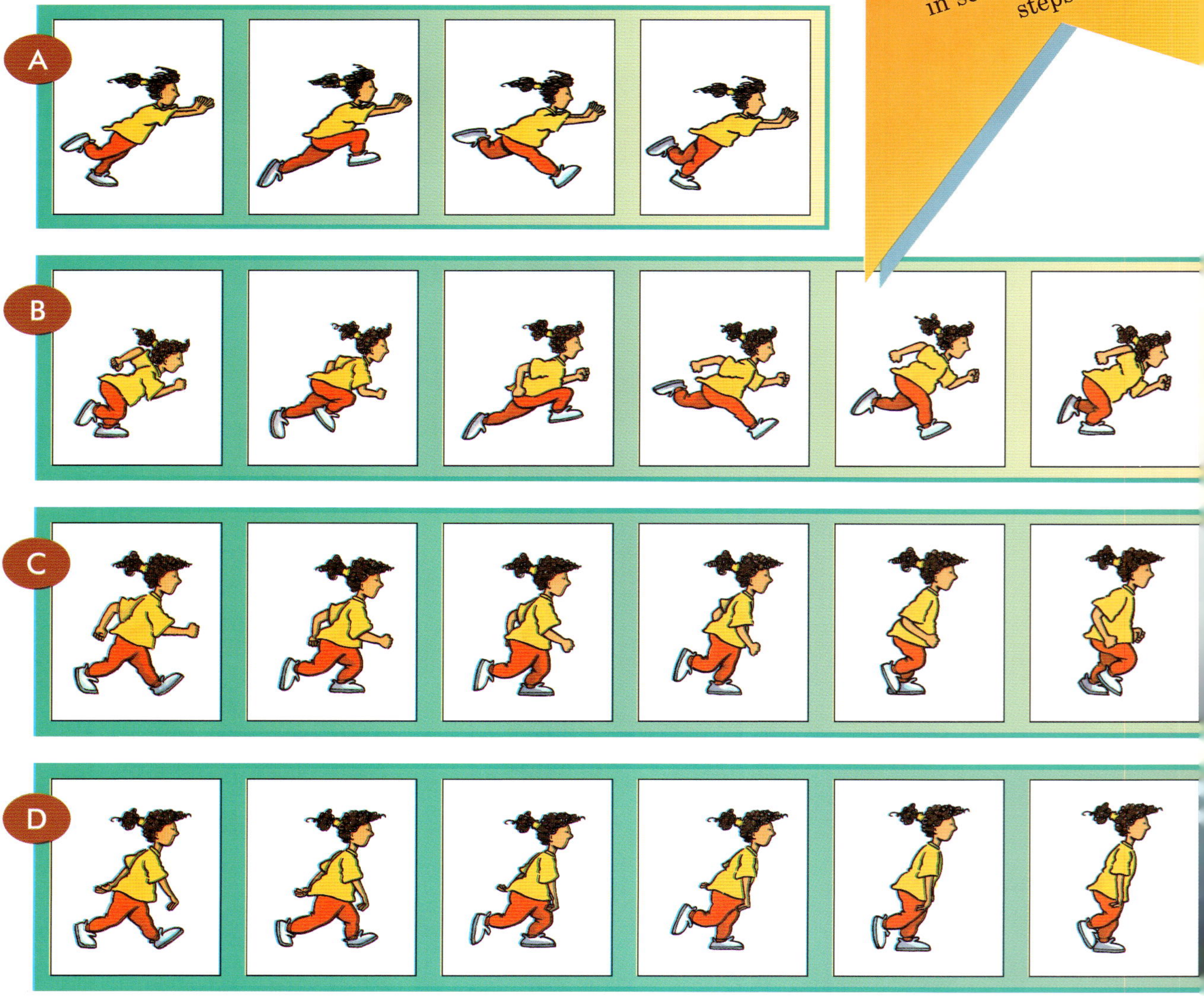

1. Use the chart on page 33 to help you figure out the number of frames and the number of feet of film used in:
   a. a one-minute cartoon
   b. a 7-minute cartoon.
2. Look closely at each set of frames. For each set, how many frames were used to show one of the girl's steps?
3. Look at the four sets of frames shown opposite. Copy and complete the chart below.

STRIP	FRAMES PER STEP	STEPS PER SECOND
A		
B		
C		
D		

A World of Movies

How many new movies have you seen in the last 12 months? Each year thousands of movies are made. Some countries make hundreds of movies, while other countries make only a few.

Many countries have been making movies since early this century. The world's first full-length movie, *The Story of the Kelly Gang,* was made in Australia in 1906.

Did you know?
India produces about 800 movies each year – many more than any other country. India has a population of almost one billion – that's 1,000 million people.

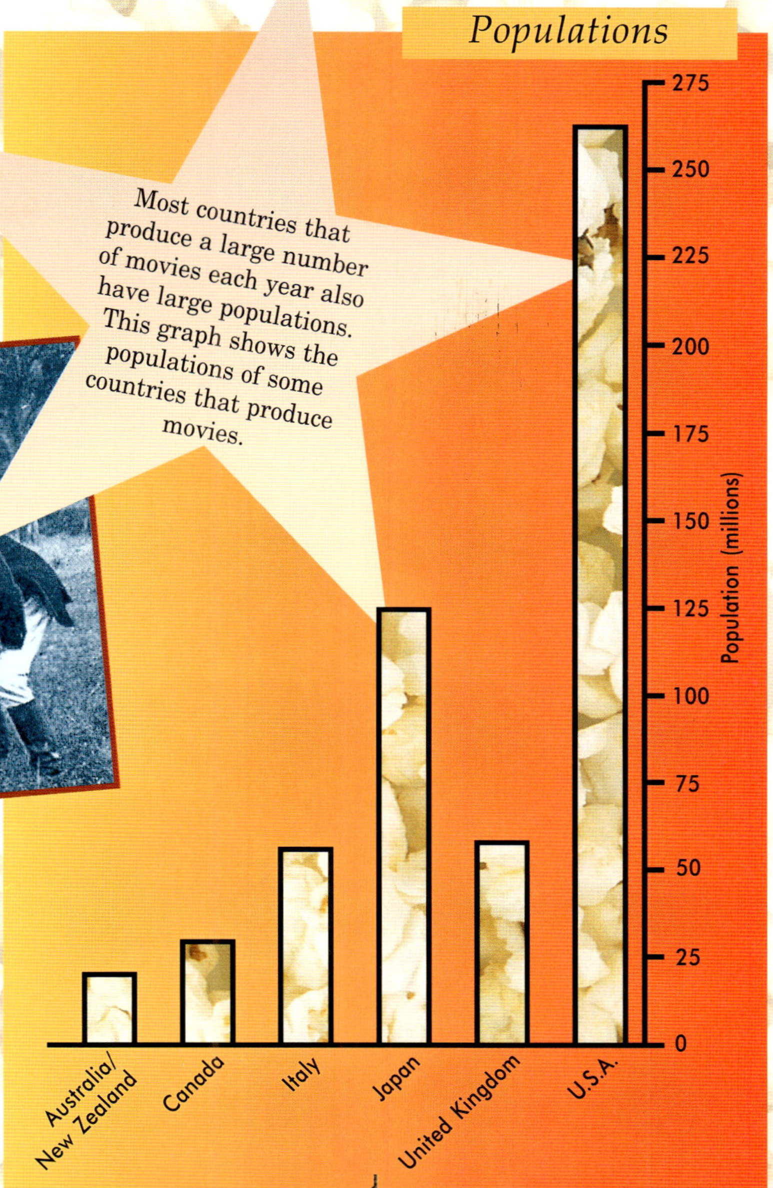

TAKE YOUR SEAT

Millions of people go to see most big-budget movies, but not all movies attract huge audiences. For this reason, many theater complexes cater for audiences of different sizes.

Most theaters seat about 250 people, but some can accommodate hundreds more.

New

CINEPLEX 8 OPENING SOON

Theater	*Seating Capacity*
1 and 2	560 each
3,4,5 and 6	240 each
7 and 8	330 each

1. Calculate the total seating capacity of the new Cineplex 8 complex. How did you figure it out?

2. Each theater in the complex has seats arranged in equal rows. Which theaters could have rows of:

a. 20 seats? **b.** 15 seats?

3. The world's largest cinema complex has 24 theaters seating 8,000 people in total. Calculate some possible seating capacities for each theater, to reach a total of 8,000 seats.

Look at the graphs on pages 36–37.

4. Choose two countries. Compare each population with the number of movies made.

5. Why do you think India was not shown on the population graph?

Theater Records

- The Radio City Music Hall, New York, is the world's largest single theater. When it opened in 1932, it had 5,945 seats. Today, it seats 5,874 people.
- The world's largest theater complex is in Belgium. It has 24 individual theaters. Each theater seats between 160 and 700 people – a total seating capacity of about 8,000 people.

Glossary

Area
The amount of surface enclosed within a boundary. Area is usually measured in square units.

Budget
An estimate of expenses or income. These are often itemized into a list.

Dimension
The size of an object measured in a particular direction, such as length, width, or height.

Pie Graph
A circle divided into sectors to show information. The area of each sector represents one piece of information.

Profit
The money left over from the income after all expenses have been paid.

Rate
A comparison between two different types of measurement involving division. For example, money and time: $50 per day.

Similar
When one shape is the same shape as another, except for its size. Similar shapes can be larger or smaller than the original shape.

Venn Diagram
A number of overlapping circles used to sort items according to their characteristics. Each circle represents a different category or group. Only those items that belong to a particular category can be placed in that circle, but some items can be shown to belong to more than one category.

Index